Summary: Scientific facts about banana slugs in the Pacific Northwest. Petroske, Jarad, author.
ISBN-13: 978-1540523877 ISBN-10: 154052387X

7 Things You Absolutely Have to Know About

BANANA SLUGS

Who it's for: This book is designed for the fluent reader or it can be a book to share!

Table of Contents

INTRODUCTION

Congratulations on choosing this book! You're about to become a banana slug expert. You'll learn how these slugs use their slime to climb up walls, and how that same slime can help them glide over sticks and stones on the forest floor. You'll also learn what banana slugs eat, where they live and why you might think twice before kissing one. Ready? Let's begin!

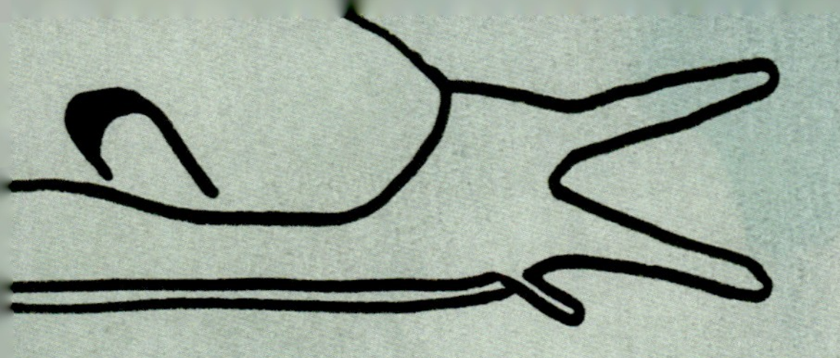

The Banana Slug

The Most Interesting Slug You'll Ever Meet

The forest is home to many creatures, and one of the most interesting and slimiest you'll meet is the banana slug of the Pacific Coast of North America. One look and it's clear why they're named after the delicious yellow fruit. While most are yellow with brown spots, banana slugs can also be completely brown and even green.

You might have guessed that banana slugs are sort of gross! And you're right. They secrete ooze, they eat rotting mushrooms and they have their stomach in their foot.

Scientists call banana slugs *Ariolimax columbianus, Ariolimax californicus or Ariolimax dolichophallus.* Phew, that's a mouthful. Banana slugs are part of the Gastropod class of animals and they count among their cousins other snails and slugs of both the underwater and land-dwelling varieties. Gastropod means "stomach foot" in Latin. That makes sense because a banana slug's mouth is also part of its single muscular foot.

BANANA SLUGS
CLASSIFIED

Ariolimax
columbianus, californicus, dolichophallus

WHERE CAN I FIND BANANA SLUGS?

Conditions need to be just right for banana slugs to thrive. Imagine tall pine, bay, oak, and redwood trees with lots of ferns and huckleberry bushes below. Last season's evergreen needles collect in the nooks and crannies between the tree roots that give shelter to banana slugs as they hide under an umbrella of tiny sorrel leaves. When you add all these incredible forest ingredients together, you have a banana slug's favorite place to live.

BANANA SLUGS LIVE ON THE COAST OF THE PACIFIC NORTHWEST

Banana slugs live as far south as the Salinas Valley of central California and on the coasts of Oregon, Washington, British Columbia, and even southeastern Alaska. These are all places where the climate stays cool and humid. The many trees, bushes, and plants provide plenty of shade and cover for banana slugs.

A banana slug would be right at home in the brown and green surroundings of the redwood forest.

Perfect slug spo

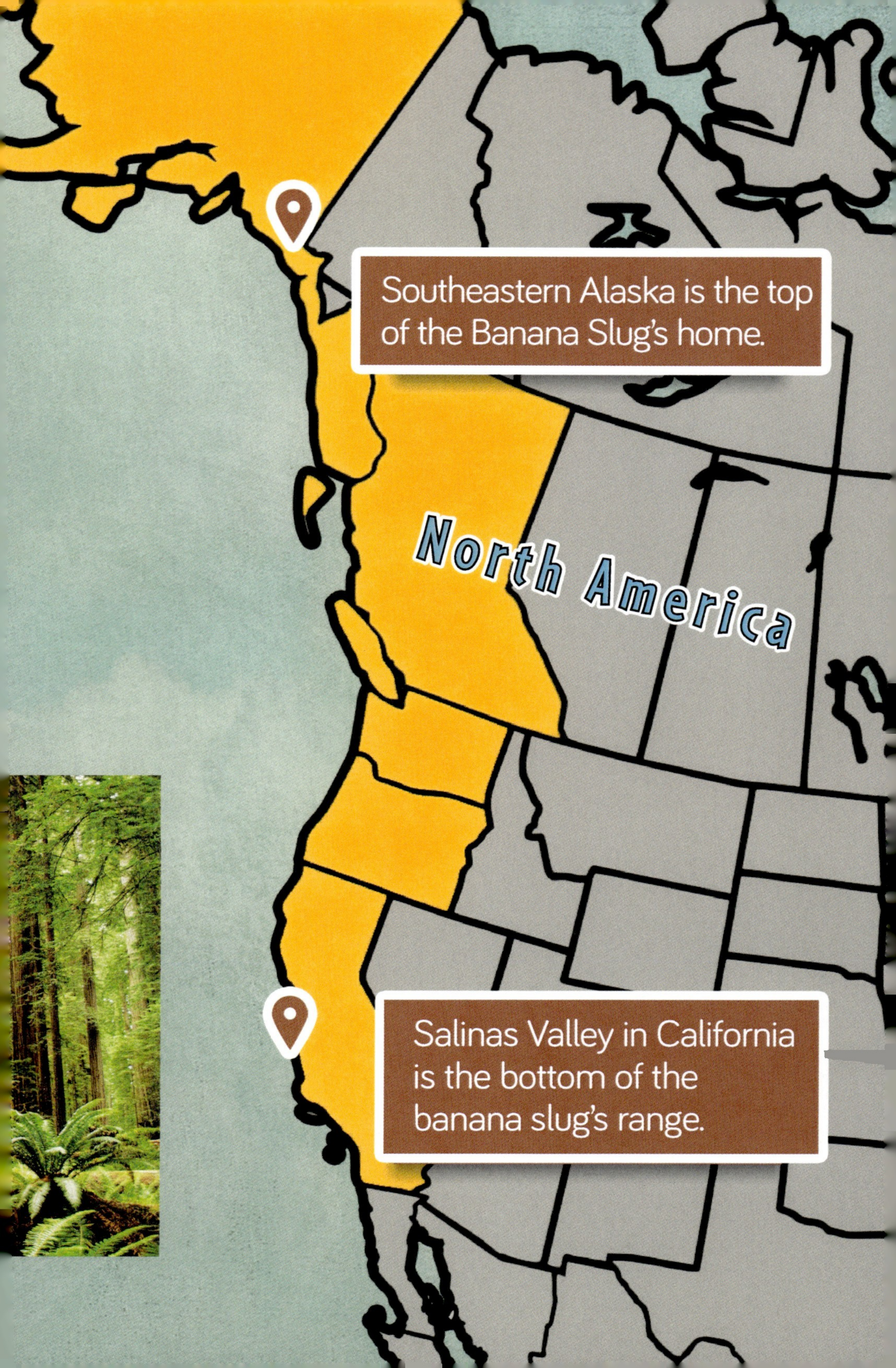
Southeastern Alaska is the top of the Banana Slug's home.
North America
Salinas Valley in California is the bottom of the banana slug's range.

WHAT'S SPECIAL ABOUT A BANANA SLUG'S BODY?

BREATHING

How do you breathe? With your nose and lungs. Banana slugs have lungs too, and breath through something called a pneumostome (new-mah-stome) that's a bit like your nostrils. Can you see the shiny skin on this banana slug? It's shiny because of a special layer of mucus slugs make around their bodies to stay hydrated and help them breathe.

VITAL STATS:

Length & Weight: Banana slugs can be up to 25 cm long and weigh nearly 115 grams.

Age Range: Incredibly, banana slugs can live up to 7 years! But most only make it 1.

Land Speed: Banana slugs are pretty slow. They can travel 17 centimeters (6.7 inches) in one minute. You can easily run 1,000 times faster than a banana slug.

SEEING & SMELLING

Like all slugs, banana slugs use tentacles to see and to do something that's kind of like smelling.

Eyestalks detect light and movement, while the smaller pair of tentacles detect scents to help banana slugs find food, detect predators, or find other slugs.

Banana slug anatomy

Denticles in a slug's radula

Munch munch munch: How do you eat? With strong teeth. But banana slugs use hundreds of tiny teeth called denticles in an organ called the radula. With the radula, banana slugs grind up the food they eat. But what are those foods?

#4 WHAT DOES A BANANA SLUG EAT?

Do you like to eat fruits and vegetables? So do banana slugs! But there are many things banana slugs will eat that you probably wouldn't. Banana slugs like to eat dead plant matter, moss, mushrooms, even animal droppings. We call animals who

eat like this detritivores (deh-trite-uh-vores). Everything in the forest is connected. After a banana slug is done eating, they leave behind a nutrient-rich excretion that helps return nutrients to the next generation of life in the forest. Because they travel—albeit slowly—banana slugs help spread fungus spores and plant seeds and that helps many of the plant and fungi species find new places to grow in the forest.

WHEN DO THEY EAT?

You might say a banana slug's favorite meal is dinner. Most of the time, the slugs prefer to do their eating at night because it's too hot during the day to stay hydrated. If the habitat is shady enough, like deep in the gullies of the redwood forest, banana slugs will snack all day long.

A banana slug uses its radula to grind up and eat wild ginger. Do you see its open pneumastome?

Could these mushrooms be the next snack for a hungry banana slug?

#5

WHO IS EATING BANANA SLUGS?

Do you think a banana slug would be good to eat? If you are a raccoon or a giant pacific salamander, then yes! These species, along with birds, snakes, and even shrews and moles will all eat banana slugs. Clever animals like raccoons will coat the slugs in leaves and dirt on the forest floor to avoid eating the slug's sticky slime.

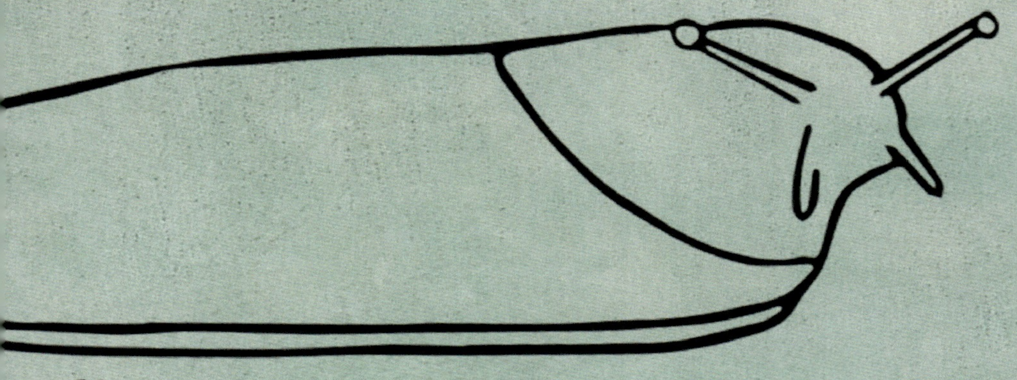

Racoon
Salamander
Shrew
Spotted Owl

HOW DO BANANA SLUGS REPRODUCE?

Once again, banana slugs have amazing habits when it comes to reproduction. For starters, banana slugs are hermaphrodites and that means each slug has both male and female reproductive organs. A banana slug can make its own fertilized eggs, but mostly they prefer to find another slug to lay eggs with.

This slug could reproduce all by itself, but they usually choose to look for a mate.

Banana slugs will circle each other as they prepare to mate.

EGG LAYERS BUT NOT EGG SITTERS

Adult banana slugs lay their eggs—up to 20 at a time—in a safe place and then leave. They don't wait around to raise their young. When young banana slugs hatch, they are born ready to take on the world.

Estivation: A Banana Slug's Super Lazy Super Power! When the temperature is too high and there's not enough moisture in the environment, a banana slug does something called estivation. In order to survive the dry conditions, the banana slug will bury itself

A slug with its clutch of eggs.

in the forest floor and cover itself in a thick layer of mucus to retain moisture. It will rest until the weather changes and it can once again be active. Next time you want to sleep in late, tell your parents you're conserving energy with estivation and you'll get up once the conditions are right!

SLIMIEST

A banana slug faces many challenges in nature: its body is mostly made of water so it can't get too hot or too dry. They don't have feet—just one big muscular foot—so moving over sticks and rocks can be hard. Sometimes their favorite food is up a tree, so

t

A BANANA SLUG'S SUPER POWER

they have to climb many times their own height to reach it. How does a banana slug meet these challenges? With incredible slug slime!

Banana slug slime is truly one of nature's miracles. The slime is a special mucus that's much like the mucus your body makes. Cells in the banana slug's body called goblet cells create tiny granules of mucus that pop open and absorb up to 100 times the granule's volume in water. A slug can actually attract water to its body using its slime.

The banana slug's slime, like all mucus, is technically a liquid crystal—which means it's half-way between a solid and liquid. This special property allows the slime to be either slippery, helping the slug glide over the rough forest floor, or sticky, helping the banana slug climb trees, stumps, and other tall surfaces.

A banana slug using its incredible slime to stick to a tree and glide over the rough surface.

A final note ...

BEWARE OF KISSING BANANA SLUGS

For reasons unknown, it's a bit of a tradition to kiss a banana slug. But you should know banana slug slime can make your lips feel numb if you try smooching one. No, kissing a banana slug won't feel like you just got done at the dentist, but you will get a tingly numb feeling in your lips or tongue.

Scientists aren't quite sure what causes the numbing, but it may have something to do with the food slugs eat. Many mushrooms and plants contain toxins and some of those toxins may be present in the slugs' slime creating the numbing properties.

Be warned, if numb lips aren't bad enough, kissing a

banana slug will leave a slimy residue on your mouth for many hours after you're done puckering up.

BANANA SLUG VOCABULARY

Arilolimax (Ari-loli-max): The scientific name for a banana slug.

Detritivore (Deh-trite-uh-vore): An animal that mostly eats deat organic matter.

Estivation (Esti-vay-shun): Prolonged sleeping or stillness during a dry period.

Gastropod (Gastro-pahd): A group of animals that includes slugs and snails. There are thousands of kinds of gastropods.

Hermaphrodite (Her-maf-roe-dite): An animal with both male and female reproductive organs.

Mollusk (Mawl-usk): A group of invertebrates that includes slugs, snails, and octopuses.

Mucus (Mew-kus): A slimy substance secreted for lubrication or protection.

Pacific Northwest: North American region where banana slugs can be found.

Pneumostome (New-ma-stome): A breathing pore found on land-dwelling slugs and snails.

Radula (Rah-doo-la): An organ, like a tongue, that mollusks use to eat.

Tentacles (Tent-a-kulls): Small flexible limbs usually near an invertebrates mouth.

Credits & Acknowledgements
Special thanks to Flickr Users:
Oregon Caves; California Dept. of Fish & Game; Kathy & sam; Brad Greenlee; A.Poulos (Iya); Amit Patel; Upupa4me; Franco Folini; Ben Stanfield; Christopher, Tania and Isabella Luna.